FROM **TREE** TO **HOUSE**

by Robin Nelson
photographs by Stephen G. Donaldson

Lerner Publications Company / Minneapolis

Lerner Publications Company
A division of Lerner Publishing Group
241 First Avenue North
Minneapolis, MN 55401 U.S.A.

Website address: www.lernerbooks.com

Library of Congress Cataloging-in-Publication Data

Nelson, Robin, 1971–
    From tree to house / by Robin Nelson ; photographs by Stephen G. Donaldson.
        p.   cm. – (Start to finish)
    Includes index.
    Contents: Loggers cut down trees – A Machine digs a hole – Builders make the basement – Builders make the floor – Builders put up the walls – Builders make the outside – Builders make the roof – Workers add wires and pipes – Workers finish the house.
    ISBN: 0–8225–1392–7 (lib. bdg. : alk. paper)
    1. House construction—Juvenile literature. 2. Building, Wooden—Juvenile literature. [1. House construction.] I. Donaldson, Stephen G. ill. II. Title. III. Series: Start to finish (Minneapolis, Minn.)
    TH4811.5.N44 2004
    690'.837—dc22   2003016544

Manufactured in the United States of America
1 2 3 4 5 6 – DP – 09 08 07 06 05 04

Table of Contents

This is my new house.

How was it built?

Loggers cut down trees.

The trees are sent to a **sawmill**. The sawmill's machines saw trees into pieces of wood. The wood is sent to builders.

A Machine digs a hole.

Machines clear the land. They
remove trees and rocks. Then
a machine digs a hole for the
house's **foundation** and
basement.

Builders make the basement.

Builders put up hollow walls called **forms**. Trucks fill the forms with a mix called **concrete**. The concrete dries and gets hard. Builders remove the forms. The dried concrete makes the basement walls.

9

Builders make the floor.

Concrete is poured for the basement floor. Builders lay long pieces of wood across the top of the basement walls. They nail thick sheets of wood onto the pieces. This is the floor of the house.

Builders put up the walls.

Builders nail pieces of wood together to make the walls. Builders push up the walls and nail them into place.

13

Builders make the outside.

Flat sheets of wood are nailed on the outside of the walls. Builders cut holes for windows and doors. Later, they add more wood and paint it.

Builders make the roof.

Builders build big triangles out of wood. The triangles are nailed to the top of the house. More pieces of wood are nailed on top of the triangles. The covered triangles make the roof.

Workers add wires and pipes.

Workers put wires in the walls for **electricity.** The house needs electricity to make things work. Plumbers put pipes in the walls. Pipes bring water to bathtubs, toilets, and sinks.

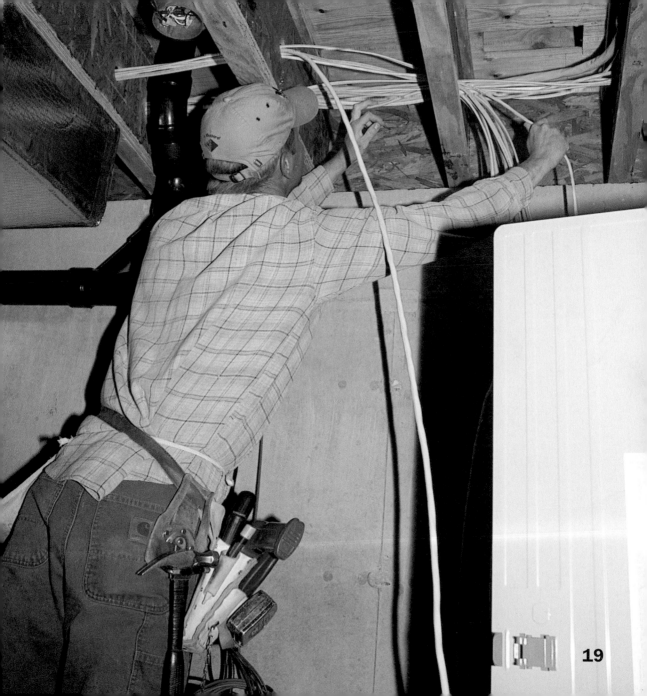

19

Workers finish the house.

Inside walls are added to make rooms. Walls also hide pipes and wires. Doors, windows, cabinets, and closets are put into the rooms. Workers cover the floors with carpet, tile, or wood.

The house is a home.

This is my new house. My family makes it a home.

HATS
SCARVES
GLOVES

MAX'S STUFF

WYLIE'S
TOYS

IAN'S
ROOM

23

Glossary

concrete (KON-kreet): a
mix of cement, sand, and
water that gets hard
when it dries

**electricity (ee-lek-TRISS-
uh-tee):** a kind of energy

forms (FORMZ): hollow
walls that shape concrete

**foundation (foun-DAY-
shuhn):** the cement part
on which a house is built

sawmill (SAW-mihl): a
place where wood is cut

Index